PRINTMAKERS THE DIRECTORY

First published in Great Britain 2006

A&C Black Publishers Ltd
38 Soho Square, London, W1D 3HB
www.acblack.com

Copyright © 2006

ISBN-10: 0-7136-7387-7
ISBN-13: 978-0-7136-7387-6

A CIP catalogue record for this book is available from the British Library.

Cover and text design by Sutchinda Rangsi Thompson
Type set in Bembo and Frutiger

Disclaimer: Information given in this book is to the authors' best knowledge correct and every effort has been made to ensure its accuracy and safety but neither the authors nor publisher can be held responsible for any resulting injury, damage or loss to either persons or property. Any further information which will assist in updating any future editions would be gratefully received.

Printed and bound in China.

PRINTMAKERS THE DIRECTORY

A&C Black • London

FOREWORD

It gives me great pleasure to introduce artists' original prints by members of the Royal Society of Painter-Printmakers. Selection for membership of the society is extremely rigorous and, as is borne out in the pages of this book, the standard is high. These are prints conceived as such from the outset by the artists. They are not photographic reproductions of paintings or drawings, but were all made directly on a plate, block, screen, stone, or computer with the sole intention of producing a print and exploiting, to the full, the particular and unique mark-making possibilities afforded by printmaking methods.

For the collector, prints are often rightly seen as a way of acquiring affordable, original works of art. An edition of 10 or even 50 prints enables the artist to charge proportionally less for each one than for a one-off piece. However the public often overlooks the reason that artists choose to make prints. When making images, artists can choose from numerous materials and tools including pencil, charcoal, many types of paint and brush. Each medium can be used in various ways to communicate the symbols and signals of touching, feeling, stroking, grasping, remembering and being moved by their subject matter, be it abstract or figurative. If these methods of mark-making can be seen as the language of art, then printmaking is a way of greatly extending that vocabulary. There are feelings that can be expressed better by a delicate lithographic wash or a bold woodblock impression than in any other way. Each printmaking technique offers the artist its own new range of marks. The artists exhibiting in this book have all experienced the delight of finding a technique that expresses just what they want to say. I hope that readers will come away infected by that delight and excited by the possibilities offered by printmaking.

ANITA KLEIN
President of the Royal Society of Painter-Printmakers

Seymour Haden RE *Wareham Bridge*, 1877, drypoint, 14.9 x 22.6 cm

Alphonse Legros RE *Death and the Woodman*, c.1875, etching, 37.3 x 27 cm

James Tissot RE *An Uninteresting Story*, 1878, etching and drypoint, 31.4 x 20.3 cm

A BRIEF HISTORY OF
THE ROYAL SOCIETY OF PAINTER-PRINTMAKERS

The Royal Society of Painter-Printmakers (RE) has been the champion of artists' printmaking from the time of its foundation in 1880 to the present day. It was established during the etching revival in nineteenth-century Britain when Samuel Palmer, Seymour Haden, his brother-in-law James McNeill Whistler and others began using the medium expressively as an original art form. London's Royal Academy of Arts showed only oil painting, sculpture and architecture at its Summer Exhibition, although engraved reproductions of members' paintings were included. Some of the engravers were elected to membership but only as associates. No artist using the medium creatively was considered eligible for election nor was his or her etching considered fine art worthy of entry. Haden could not persuade the Academy to change its policy so, in protest, he established the Society of Painter-Etchers to promote original printmaking, garnering wide national support, with backing from European and American artist-printmakers working in Britain.

Frank Short RE *The Tide Ebbs, Putney Bridge*, 1885, mezzotint 125 x 181 mm

Walter Sickert RE *Louie*, 1884, etching, 22.8 x 22.5 cm

Malcolm Osborne RE
Maggie, Study of a Girl's Head,
1905, etching, 12.7 x 6.5 cm

Eminent men and women were elected to membership, mounting the Society's first show in London at the Hanover Gallery in 1881. Annual exhibitions were subsequently held at the Fine Art Society in Bond Street, with provincial exhibitions in Liverpool at the Walker Art Gallery in 1884 and in Derby in 1886. That year, an invitation was accepted to share premises with the Royal Watercolour Society (RWS) at their gallery in Pall Mall East, when Queen Victoria recognised the Society's aims by granting the title "Royal". As its prestige grew, King Edward VII issued in 1911 a Royal Charter for the Society to confer diplomas on associates and fellows. Relations with the Royal Academy improved when Sir Frank Short RA became the second President of the RE and, since 1915, Academy Presidents have been elected Honorary Fellows of the Society.

Redevelopment of Trafalgar Square prompted the RE and RWS to move to the West End in 1938, where they remained at 20 Conduit Street until the lease expired on the RWS Galleries in 1980. A permanent home was finally established in Blackfriars, at the Bankside Gallery, with the intention of having a full programme of exhibitions to promote works on paper and to function as an educational charity. Gradually extending the scope of its exhibited printmaking from intaglio to relief, the society renamed itself the Royal Society of Painter-Etchers and Engravers in the 1920s. In 1992 it embraced all forms of original printmaking and thereupon restyled itself again as the Royal Society of Painter-Printmakers, keeping the short form of RE, as it has come to be known.

In the modern world of advancing technical sophistication in printing techniques, the Society continues to embrace innovation whilst fostering the creative production of graphic art as quite different from the commercial reproduction of existing

Gwendolen Raverat RE *The Gooseherd*, 1919, wood engraving, 10.1 x19.1 cm

Laura Knight RE
Putting On Tights, 1926
etching, 20.2 x 17.6 cm

Graham Sutherland RE *No.49*, 1924, etching, 17.7 x 25.2 cm

Agnes Miller Parker RE *The Challenge*, 1934, wood engraving, 14.2 x 16.3 cm

Gertrude Hermes RE *Self Portrait, No. 19,*
1949, linocut, 20.4 x 15.3 cm

Monica Poole RE Shell, 1967,
wood engraving 23 x 16.8 cm

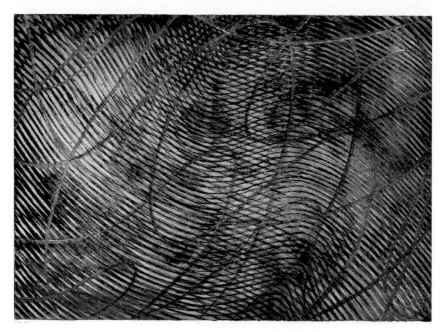

Stanley Hayter RE *Lunar Rhythm*, 1967, etching and aquatint, 34.7 x 48.3 cm

Valerie Thornton RE *Montacute House*, 1957, etching and aquatint, 28.6 x 51.8 cm

Michael Rothenstein RE *Subway*, 1985, woodcut, 56.3 x 75.8 cm

works. It pursues the aim set out in its charter to provide the public with an original form of art at a reasonable cost and to demonstrate the best examples of artists' printmaking in members' exhibitions. The Society attracts a growing international membership, electing the most accomplished printmakers working today, who share a wish to promote this uniquely expressive and democratic of all forms of art.

JOSEPH WINKELMAN PPRE

Note:

The prints illustrating this brief history were all made by RE members since the Society's foundation. The works are selected from amongst the RE's extensive Diploma Collection of members' work, which may be viewed on request at the Ashmolean Museum, Oxford.

Photographs of RE Diploma Collection prints © Nigel Swift 2005

Hilary Adair studied painting and printmaking at Brighton and St Martin's Colleges of Art. She lives on Exmoor in Somerset and also has close connections with Greece, her first solo exhibition being presented by the British Council in Athens. She has carried out commissions for prints and paintings and has exhibited widely. Her work is held in many public and private collections.

Koskina, 1999, etching, 54 x 90 cm

Halse Cottage
Winsford
Minehead
Somerset TA24 7JE

tel.   **01643 851 238**
email   hilaryadair@halsecottage.fsnet.co.uk
web   www.hilaryadair.co.uk

TATE **ADAMS** RE

Tate Adams studied engraving with Gertrude Hermes in London. In 1951 he went to Australia and taught printmaking in Melbourne. He became director of the Crossley Print Gallery and founded the Lyre Bird press from which he has published 35 artists' books to date. His work is in the Tate Gallery, London; the Metropolitan Museum of Modern Art, New York; the National Gallery of Australia; and in other collections.

Palms, wood engraving, 38 x 38 cm

© Andrew Rankin Photography

401/30 Palmer Street
Townsville
North Queensland
Australia 4810

2

George Adamson, born in 1913, studied at Wigan and Liverpool Schools of Art and at the latter specialised in engraving under Geoffrey Wedgwood RE. After a long career as a book illustrator and cartoonist, he returned to printmaking in the late 1970s and worked until he was over eighty. He died in March, 2005. His work is in several public collections including those of the V&A, the British Museum and the New York Public Library.

Wine Steward, 1937, soft ground etching and aquatint, 31.3 x 22.8 cm

George W Adamson Archive
c/o 90 Hertford Street
Cambridge CB4 3AQ

tel.   01223 313 717
web   www.georgewadamson.com

TREVOR **ALLEN-ABBOTT** RE

Trevor Allen-Abbott studied at Camberwell School of Art and taught printmaking there and at Goldsmiths' and Ravensbourne Colleges. One-man shows have been held at London's Serpentine Gallery and elsewhere, and he has participated in exhibitions in Cracow, Milan, the USA and Ireland. His work was shown in the BBC television series 'The Artist in Print' and is represented in public and private collections in Britain, USA, Canada and Europe.

Blue Shade, 1996, linocut, 40 x 53 cm

Tor Barn
Lower Godney
Wells
Somerset BA5 1RZ

tel.   01458 832 996

Jim Anderson is a printmaker, painter and mosaic artist. He studied at Oxford University and London's Central St Martin's School. His prints – which combine satire and surrealism – draw on a wide range of techniques, and use of recycled materials is paramount. He exhibits internationally and, in 2000, his book *Handmade Prints*, co-written with Anne Desmet, was published by A&C Black.

Book Lover (Man of Letters), 2002, mixed media on handmade paper, 103 x 51 cm

70 Bickerton Road
London N19 5JS

tel.   **020 7263 4651**
email   **jimpanzee2001@yahoo.co.uk**
web   **www.jimpanzee.co.uk**

TED ATKINSON RE FRBS

Ted Atkinson trained as a sculptor at Liverpool College of Art and the Slade School, later studying printmaking at the Central School, London. Representative prints can be found in numerous public collections including those of the Ashmolean Museum, Oxford, the Fitzwilliam Museum, Cambridge, and the Stadt-Wetzlar Museum, Germany. A book, *Ted Atkinson, Etchings*, with sixty full-page illustrations, was published in 1997.

Night Geometry 6, 2003, computer aided etching, 28 x 25 cm

4 De Vere Close
Wivenhoe
Essex CO7 9AX

tel.   01206 825 887

Mychael Barratt was born in Canada and studied printmaking at the Nova Scotia College of Art, Halifax, and Central St. Martin's, London. His narrative work draws heavily on his interest in literature, theatre and art history. He exhibits widely in many countries and is a commissioned artist for Shakespeare's Globe Theatre in London.

Private View, 2004, etching and aquatint, 49 x 66 cm

23 Blue Anchor Lane
London SE16 3UL

tel.   020 7232 1916
web   www.mychaelbarratt.com

JO BARRY RE

Jo Barry lives and works in the New Forest and her etchings, drawings and water colours are inspired by this ancient landscape. Her work can be seen throughout the UK and in many galleries on the east coast of America.

Silent Touches of Time, 2004, etching, 17.5 x 19.5 cm

Wing House
Hightown Hill
Ringwood
Hampshire BH24 3HG

tel.   01425 475 474
email   Barry4lee@aol.com

CHARLES **BARTLETT** PPRWS RE ARCA

Charles Bartlett studied at Eastbourne School of Art and the Royal College of Art. He did military service in the Royal Signals (1942–1947) and subsequently studied painting and printmaking. He was Senior Lecturer at Harrow College of Art, 1960–1970. His work is based principally on landscape and the coast of East Anglia. Examples of his work are in the collections of the Victoria and Albert Museum; the British Museum; the Albertina, Vienna; and the National Gallery of Western Australia.

Yacht Harbour, 1980, etching and aquatint, 40.6 x 45.7 cm

St Andrews House
Chapel Road
Fingringhoe
Colchester CO5 7BG

RICHARD **BAWDEN** RWS RE

Richard Bawden is a painter, printmaker and designer. He works in lino and etching, occasionally combining the two. He has had over 50 one-man exhibitions of prints and watercolours and commissions for engraved glass, murals mosaics, textiles and cast iron furniture. He studied at the Royal College of Arts, has taught part-time at London and provincial art schools, and is a past chairman of Gainsborough's House Print Workshop, Suffolk.

Fireside, 2004, linocut, 41 x 48 cm

72 Benton Street
Hadleigh
Suffolk IP7 5AT

tel.+fax **01473 823 193**
email   **h.r_bawden@btopenworld.com**

James Beale studied in Gloucestershire and at Chelsea School of Art between 1968 and 1973. His work is a response to landscape. He exhibits regularly in the UK, Europe, Japan and the USA. His work is in public and private collections in the UK and abroad. He is currently senior lecturer and Head of the Printmaking Department at the University of Gloucestershire.

The Horizon, etching, 46 x 64 cm

7 Charlecote Corner
Bishops Cleeve
Cheltenham
Gloucestershire
GL52 8HG

tel.   01242 679281
email   jamesbeale@hotmail.com

JUNE **BERRY** RWS RWA NEAC RE

June Berry studied at the Slade School. She was Vice-President of the RWS, 2001–2004. Her work is in the collections of the Government, the Ashmolean Museum, the Royal West of England Academy, the National Museum of Wales, H.M. the Queen, the Graphotek (Berlin), Union of Bibliophiles (Moscow), Wooster Museum and the Hope College Museum (Ohio), Oldham Art Gallery, Kettering Art Gallery and the Universities of Cardiff, Southampton, Sussex and Ulster.

Two Sisters in the Garden, 1995, drypoint and hand colour, 43 x 36 cm

tel.   **020 8658 0351**

Jeremy Blighton was trained as an intaglio printer and ultimately became studio manager at Thomas Ross and Son, London, 1978–1993. He was Master Printer at London Contemporary Art from 1993–2004 and has wide experience of printing historic and modern plates. A painter and printmaker in his own right, he was elected to the RE in 2004 and has exhibited in Britain and abroad, notably at the RA and in Poland.

The Long Man of Wilmington, 2004, etching and aquatint, 15 x 25.5 cm

6 Juer Street
Battersea
London SW11 4RF

tel.   020 7228 1366

MANDY **BONNELL** RE

Mandy Bonnell studied at the Royal College of Art and has exhibited regularly in the UK and Kenya. She is a producer of book art and her work is held at the Eagle Gallery, London, as well as in public collections world-wide, notably the Yale Centre for British Art, USA.

Coral Fish, 2000, reprinted 2005, wood engraving, 23 x 20 cm

Basement Flat
11a Grange Road
London SE1 3BE

email   mand@dbpress.demon.co.uk

James **Bostock** studied at Medway and the Royal College of Art in the late 1930s. He was Vice-Principal, West of England College of Art (1965–1970) and Academic Development Officer, Bristol Polytechnic (1970–1978). His prints are in the British Museum, the V&A, the British Council and numerous other collections. The last decade has for him been productive, with a retrospective exhibition of prints at the Twentieth Century Gallery, London.

The Mirror, 1951, wood engraving, 15.2 x 15.2 cm

c/o Hilary Chapman
Twentieth Century Gallery
821 Fulham Road
London SW6

tel.   020 7384 1334
web   www.hilarychapmanfineprints.co.uk

JAMES **BOYD-BRENT** ARE

James Boyd–Brent studied at Central St Martin's in London and at the University of Minnesota. He lives and works in Minnesota. His prints and watercolours are in public and private collections in the US and the UK.

The Mississippi, St.Paul, 2002, intaglio, 60 x 90 cm

240 McNeal Hall
1985 Buford Avenue
University of Minnesota
St Paul
MN 55108
USA

email   jboydbre@che.umn.edu

Simon Brett has been a leading figure in the revival of wood engraving, as an author and curator and as Chairman of the Society of Wood Engravers (1986–1992). Trained at St Martin's School of Art, London, he learned engraving from Clifford Webb. He travelled as a painter before specialising in engraving and has illustrated over forty books, but finds free printmaking of increasing importance.

Run for Your Country, Run for Your Life, 2003, wood engraving, 20.2 x 25.3 cm

12 Blowhorn Street
Marlborough
Wilts SN8 1BT

tel.   01672 512 905
web   www.simonbrett-woodengraver.co.uk

PENNY **BREWILL** ARE

Penny Brewill studied Fine Art at the University of the West of England and Postgraduate Printmaking at the Slade School of Fine Art. She has taught at Winchester School of Art and currently teaches at the Curwen Print Study Centre, Cambridge, and has previously carried out consultancy work for John Purcell Paper. She exhibits her work worldwide and has won several major awards including The Boise Scholarship. She has undertaken commissions from BBC TV, Nokia and McDonald's.

Good Fences Make Good Neighbours, 2004, etching, 56 x 38 cm

9 Woodriffe Road
Leytonstone
London E11 1AH

tel.   020 8257 3955
email   penny.brewill@ntlworld.com

Jane Brigstock studied Fine Art (Painting) at Maidstone School of Art and Printmaking at Chelsea. She has been guest artist at California College of Art and Crafts, Oakland, USA. She is a printmaker who also works in pastel and water-colour. She teaches and exhibits nationally and internationally.

Leaves and Stars, 1993, screenprint, 63.5 x 89 cm

© Jon-Firth

76 West Hill Road
St Leonards On Sea
East Sussex TN38 0NE

tel.   01424 420 145
email   jane.brigstock@btinternet.com

JANET **BROOKE** RE

Janet Brooke studied at Birmingham and Brighton Colleges of Art. She exhibits widely both in Britain and abroad and has works in many private and public collections including the Museum of London. She is a founder-member and exhibitions' co-ordinator of East London Printmakers, and has worked extensively in the area of promoting printmaking within education.

Missing on the Hackney Road, screenprint, 56 x 76.2 cm

61 Brokesley Street
London E3 4QJ

email   mail@janetbrooke.com
web   www.janetbrooke.com

John Brunsdon studied at Cheltenham School of Art and at the Royal College of Art, where the innovative approach of Julian Trevelyan to colour etching was a formative influence and inspired Brunsdon to produce images with remarkable colour combinations owing little to direct observation of nature. His main interest is in man's effect upon the landscape. The V&A, Tate Gallery, New York's Museum of Modern Art and many other major institutions hold examples of his work.

Sussex Coastline with Martello Tower, 2004, etching, 45.7 x 61 cm

© Richard Denyer

Old Fire Station
Church Street
Stradbroke
Eye
Suffolk IP21 5HG

tel.   01379 384 647 (Studio)
tel.   01379 388 236 (Home)

CORINNA **BUTTON** ARE

Corinna Button studied painting at Leeds College of Art and printmaking at Croydon School of Art. Her work explores the hidden tensions and dramas of daily life and relationships. She exhibits regularly in galleries throughout Britain, including various major international shows. Button has also exhibited in Korea, New Zealand and across Europe.

Rush Hour, 2003, carborundum and collagraph, 56 x 61 cm

© Fraser Marr Photography

email   corinna@corinnabutton.com
web   www.corinnabutton.com

Neil Canning was born in Oxfordshire in 1960 and studied in the studio of a professional painter in preference to enrolling at an art school. He worked on his first set of prints at Advanced Graphics London in 1995 and has subsequently made many monoprints and woodblock prints. Canning is currently based in St Ives.

Fusion VI, 2003, screenprint with woodblock, 47 x 58.5 cm

c/o Advanced Graphics London
32 Long Lane
London SE1 4AY

tel.   020 7407 2055
web   www.advancedgraphics.co.uk

DAVID L **CARPANINI** PPRE HON RWS RWA NEAC

David Carpanini was born in Glamorgan in 1946. He studied at Gloucestershire College of Art, Cheltenham; the Royal College of Art; and the University of Reading. He was President of the Royal Society of Painter-Printmakers, 1995–2003, and was formerly Professor of Art at the University of Wolverhampton. His work has been the subject of three television documentaries and is held in numerous public and private collections in the UK, Europe, Canada, Saudi Arabia and Australia.

A Bit on the Wild Side, etching, 32.5 x 45 cm

© Bernard Mitchell

Fernlea
145 Rugby Road
Milverton
Leamington Spa
Warwickshire CV32 6DJ

tel.   01926 430 658

24

Daphne Casdagli, of Greek origin, was born in Cairo and studied at the Ecole des Beaux-Arts de Versailles and the Royal College of Art. She taught for 25 years in London art colleges, including the City and Guilds Art School where she became Head of Printmaking, whilst working as a painter-printmaker. She has exhibited extensively in the UK and also in Greece.

The Trundle, Goodwood, 2003, drypoint and collagraph, 41 x 61 cm

email **daphnecasdagli@onetel.com**

JEFF **CLARKE** RE

Jeff Clarke studied at Brighton and was British Institution Fund Scholar and Rome Scholar, 1956–1958. He lives in Oxford, where he has had some 15 solo exhibitions at Bear Lane Gallery, Modern Art Oxford, Oxford Gallery and Christ Church Picture Gallery. He has shown regularly at mixed exhibitions: Royal Academy; London Original Print Fair; and Mall Galleries. His work is in the collections of the Ashmolean Museum and the Universities of Oxford, Cambridge and Reading.

Old Enamel Jug and Windfalls, 2002, etching and aquatint, 15.5 x 22 cm

17 Newton Road
Oxford OX1 4PT

tel.   01865 241 634

Louise Clarke is a graduate of Central St. Martin's and The Royal College of Art. She has exhibited widely in solo shows and group exhibitions including the Jerwood Drawing Prize (2001) and Stray, a drawing exhibition in Chicago (2004). Besides working as an artist, Clarke is an experienced exhibition curator. She has also worked within arts management, specifically for the annual launch events of the Big Draw campaign.

Feather Light, 2003, screenprint, 61 x 50.8 cm, study for larger 2.4 x 1.5 m piece

tel. **07970 867 032**
email **lou.rol@btinternet.com**

KATIE **CLEMSON** RE

Katie Clemson is Australian. She trained in London at the Central School of Art in the 1970s and has an MA in Public Art from the University of East London. White Gum Press is her studio, specialising in linocuts, relief prints and typography. She exhibits at the Redfern Gallery in London and at Beaver Galleries in Canberra, Australia.

Boatshed 22, 2004, linocut, 32.5 x 25 cm

3 St Peter's Wharf
Chiswick Mall
London W6 9UD

email   katie@whitegum.com
web   www.whitegum.com

Deanne Coleborn gained a Royal Exhibition enabling her to study at the Royal College of Art. She is a member of the Greenwich Printmakers group. Her work is mainly figurative and reflects a strong sense of colour. Her themes are derived largely from the theatre, dance, music and mythology. She exhibits widely in Britain.

Dancing , 2002, etching, 20.3 x 15.2 cm

Downe Hall Farm
Downe
Kent BR6 7LF

tel.   01689 853 789

EILEEN **COOPER** RE RA

Eileen Cooper studied at Goldsmiths College and the Royal College of Art. She has exhibited widely in Britain and abroad and was elected to membership of the Royal Academy in 2000. She is represented by Art First Gallery and her work is in many private and public collections including those of the Arts Council, the British Museum, the V&A, Unilever plc and Yale University's Walpole Library.

High Kicks, 2002, linocut, 32 x 14 cm

Art First Gallery
9 Cork Street
London W1

email   eileen@eileencooper.co.uk
web   www.artfirst.co.uk

Alistair Crawford studied at the Glasgow School of Art and Aberdeen College of Education. In 1990 he became the first Professor of Art of the University of Wales. He is a painter, printmaker and art historian. His work is exhibited internationally and is represented in public, corporate and private collections worldwide. He has published over 120 books and articles in various languages and has curated exhibitions in Britain, Europe and the USA.

Rain cloud, etching and aquatint, 50.8 x 66 cm

Brynawel
Comins Coch
Aberystwyth SY23 3BD
Wales

email   alc@aber.ac.uk

PETER **DAGLISH** RE

Peter Daglish is a painter–printmaker and jazz saxophonist. His themes are city life, poetry and jazz. He has held 38 solo exhibitions in Britain, France, Germany, Canada, India, Pakistan and St Lucia. His work is in the collections of the Tate Gallery; the V&A; the British Council; Vancouver Art Gallery; Glenbow Museum, Calgary; Gilkey Collection, Portland Museum, USA, Tamarind Collection, University of New Mexico, USA, Museu de Arte de Macau; British Council, India; and the Musée d' Art Contemporain, Montreal. He has taught at London's Slade School and Chelsea for over 20 years.

Konark, 1997, linocut in colour, 50.8 x 39.2 cm

22 Mayfield Road
Wimbledon
London SW19 3NF

tel. + fax. 020 8543 2607

Claire Dalby studied engraving and lettering at the City and Guilds of London Art School. In addition to wood engraving, reflecting her interest in fine tonal gradations, she paints watercolour still lifes and landscapes and makes scientific illustrations (fungi, lichens, and other plants). She has exhibited widely, including frequently at RA Summer Shows, and her work has been purchased for numerous public and private collections worldwide.

Drummond Castle, 2000, wood engraving, 9.7 x 12.7 cm

2 West Park
Stanley
Perthshire PH1 4QU
Scotland

tel.   01738 827 222

HARVEY **DANIELS** HON. RE

Harvey Daniels studied at the Slade School. He is a Past Chairman of the Printmakers' Council and is the author of several books on printmaking. Examples of his work are in the Victoria and Albert Museum, Tate Britain, the Museum of Modern Art MOMA, New York and other major institutions. In 2003 he was awarded a Royal Watercolour Society prize and his most recent exhibition was in Aachen, Germany, in 2004.

Dove Grey, 2000, coloured woodcut, 130 x 62 cm

© Younas Mohammed

95 Springfield Road
Brighton BN1 6DH

96 Chemin de la
Caladette
30350 Lézan
France

tel. **01273 556 307**
email judy@harveydaniels.fsnet.co.uk
web www.harveydaniels.com

tel. **0033 4 66 83 25 48**

34

Anthony Dawson has been a fellow of the Royal Society of Painter-Printmakers since 1991 and shows drawings and prints regularly at the John Davies Gallery (Stow), Thompson's Gallery (Aldeburgh) and the Royal West of England Academy. His work is represented in the collections of the Ashmolean Museum and Hereford City Museum, as well as in many private collections.

Blue Boat, Walberswick, 2001, etching, 9.8 x 14.6 cm

Cider Press
Cotheridge Lane
Eckington
Pershore
Worcestershire WR10 3BA

tel.   01386 750757

ANNE DESMET RE SWE

Anne Desmet's major retrospective, 'Towers and Transformations', opened at Oxford's Ashmolean Museum in 1998. Many international exhibitions and awards include a British School at Rome Scholarship (1989/90). Her work is represented in significant museum collections and commissions include wood engravings for the British Museum, British Library and Sotheby's. She is Editor of *Printmaking Today* magazine and is also co-author of the book *Handmade Prints* (A&C Black).

Babel/Vesuvius, 2002, linocut, wood engraving, flexograph, collage, 86.3 x 78.7 cm

c/o Hart Gallery
113 Upper Street
Islington
London N1 1QN

tel.   020 7704 1131
email   info@hartgallery.co.uk
web   www.hartgallery.co.uk

Holly Downing was born in California and lived in the UK from 1974–1980. She studied at the University of California, Santa Cruz, and the Royal College of Art. Her mezzotints are exhibited internationally and have received numerous awards. Examples are in the V&A, the Ashmolean Museum, the Arts Council collection and museums in Scotland, France, Poland, Japan, Norway, Korea, the Philippines and the US.

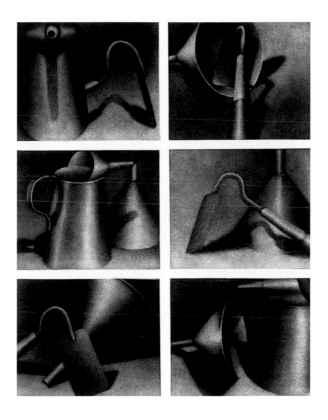

La Familia, 1998, mezzotint, 40 x 30.4 cm

5929 Fredricks Road
Sebastopol
CA 95472
USA

tel.   (707) 823 6286
email   hollydowning@comcast.net

JOHN DUFFIN RE

John Duffin is a painter and printmaker whose subject matter is the urban world; he depicts town and city architecture and the lives of their inhabitants. His recent prints have concentrated on views of London, often from unusual and dynamic angles, showing the city as a vast metropolis populated by marginal lost figures.

Double Life, 2001, etching, 38 x 25 cm

13 Marsala Road
London SE13 7AA

tel.   020 8314 1125
email   j.duffin66@yahoo.co.uk

Stuart Duffin studied at Gray's School of Art, Aberdeen, graduating in 1982 in Fine Art (Printmaking). Working mainly in mezzotint and etching, he has been a member of staff at the Glasgow Print Studio since 1984. He has had solo exhibitions in the USA, New Zealand, Israel and the UK. Awards include Scottish Arts Council and British Council grants for study and practice overseas.

Strike the Match, Stand Well Back, 2002, mezzotint, 21 x 54 cm

40 Cromarty Avenue
Glasgow G43 2HG
Scotland

email   info@stuartduffin.com
web   www.stuartduffin.com

MEG **DUTTON** RE

Meg Dutton lives and works in London. She studied fine art at Chelsea School of Art and became a member of the Royal Society of Painter-Printmakers in 2001. She regularly exhibits work in London and at other galleries throughout Britain and in Europe. She has work in various public and private collections.

Afternoon Shadows, 2002, etching and watercolour, 61 x 49 cm

tel.   **020 8673 0026**

Anthony Dyson studied at Leeds College of Art and did doctoral research in the history of printmaking at the Courtauld Institute, University of London. In 1987 he established his Black Star Press; there, he has produced many print editions for prominent artists and publishers and for national museums. His own prints are in private and public collections in Britain, Belgium, Norway and the USA.

Into the Attic, 2004, etching and aquatint, 39 x 29.5 cm

© Tamra Cave

Black Star Press
61 Hampton Road
Teddington TW11 0LA

tel.   **020 8977 2607**
email   **dysons@orange.net**

EDWINA **ELLIS** RE SWE

Edwina Ellis's engravings may be seen at Godfrey and Watt, Harrogate, and the Redfern Gallery, London. She has works in the collections of the Victoria and Albert Museum, the Ashmolean Museum, the Museum of London, the London Transport Museum and the State Library and Art Gallery both of Sydney, Australia. She designed the new British one-pound coin series in linocut.

Perched, 2001, three-block relief engraving on acetal resin, 12.7 x 10 cm

© Royal Mint photographer

Rhydgoch
Ystrad Meurig
Ceredigion SY25 6AJ
Wales

email   info@edwinaellis.co.uk
web   www.edwinaellis.co.uk

Fernando Feijoo gained his BA in Illustration, then received an AHRB scholarship to study for his MA in Fine Art (Printmaking). He won the Gwen May Trust award from the RE and also won the FBPA 'Design a Book of Fables' competition. He now works at the Curwen Studio assisting artists to produce hand-made lithographic prints. He organises his own exhibitions regularly.

Busted, 2002, linocut, 30 x 42 cm

4 Aylestone Road
Cambridge CB4 1HF

tel.   07946 743 730
email   feijoo@talk21.com
web   www.fernandofeijoo.com

PETER **FORD** RE RWA

Peter Ford has had a long career as an artist and exhibition organiser. In 1994 he added papermaking to his repertoire and is an active member of the International Association of Papermakers and Artists. He has taken part in many international exhibitions, winning over 20 prizes. His work is in many public collections, including those of the Victoria and Albert Museum, the National Art Library and the Artists' Books Collection at Tate Britain.

Tundra, 2003, relief print on artist's own paper, 108 x 80 cm

Off-Centre Gallery
13 Cotswold Road
Windmill Hill
Bedminster
Bristol BS3 4NX

email   **offcentre@lineone.net**

TREVOR **FRANKLAND** PRWS RE RBA HON. RI

Trevor Frankland studied painting at The Royal Academy Schools. He has had many solo exhibitions, several assisted by Arts Council England. He is President of The Royal Watercolour Society and a member of The London Group and other art societies. The recipient of many awards, his work is in various public and private collections. A domestic landscape constructed by him has been the subject of four short films for television.

Venetian Interior, 2003, linocut, 60 x 48 cm

13 Spencer Road
London SW18 2SP

tel.   020 7228 4321
web   www.trevorfrankland.co.uk

ANTHONY **FROST** ARE

Anthony Frost was born in St Ives in 1951 and studied at Cardiff College of Art. He has made editions with print workshops such as Gresham Studios and Advanced Graphics London, usually in silkscreen and woodblock. Currently based in St Ives, Frost has exhibited across the UK and Ireland, showing paintings and monoprints.

Walking into Blue, 2001, screenprint with woodblock, 50 x 53.5 cm

c/o Advanced Graphics London
32 Long Lane
London SE1 4AY

tel.   020 7407 2055
web   www.advancedgraphics.co.uk

Colin Gale co-founded (with Melvyn Petterson in 1992) the Artichoke Print Workshop, London, a leading open-access studio. Other professional artists including Tom Phillips, Andrew Logan, Terry Frost and Gary Hume have availed themselves of his expertise in the production of editions of original prints. Gale is the author of *Etching and Photopolymer Intaglio*, A&C Black, 2005.

Blue Carbon, 2000, etching, 90 x 50 cm

Artichoke Print Workshop
Bizspace S1
245A Coldharbour Lane
London SW9 8RR

email   colingale@btinternet.com
web   www.printbin.co.uk

DAVID **GLUCK** ARCA RWS NEAC RE

David Gluck studied at Wakefield, Leeds and the Royal College of Art. He taught printmaking at the Royal Academy Schools and Goldsmiths College and was Head of Printmaking and Fine Art Course Director at Central St. Martin's. The London Group is among the professional bodies of which he is a member. He works mainly in watercolour and etching and shows extensively throughout Britain, including at the Royal Academy Summer exhibitions.

Late Evening Gondolas, Venice, 2004, etching and aquatint, 60 x 66 cm

49 Holmdene Avenue
Herne Hill
London SE24 9LB

tel.   020 7274 5818

Carmen Gracia, born in Argentina, studied Fine Art at the Escuela de Bellas Artes de Mendoza and furthered her training at 'Atelier 17' in Paris (1960) and the Slade School in London (1965). Participating in more than 300 group and 30 solo exhibitions world-wide, Gracia's prints and paintings adorn many public collections including those of the Victoria and Albert Museum and the British Museum.

Down the Tajo Banajo Road, 1998, etching, 71 x 56 cm

65 Westover Road
High Wycombe
Bucks HP13 5HX

tel.   **01494 528 480**
web   **www.carmengracia.com**

TERRY **GREAVES** RE

Terry Greaves studied at the Central School of Art and Design. In 1963 he was awarded the Prix de Rome in Engraving. He was elected a fellow of the Royal Society of Painter-Printmakers in 1990. He exhibits widely and has works in public and private collections in the UK, Europe, North America, Australia and Japan. He has taught printmaking and painting throughout his working life.

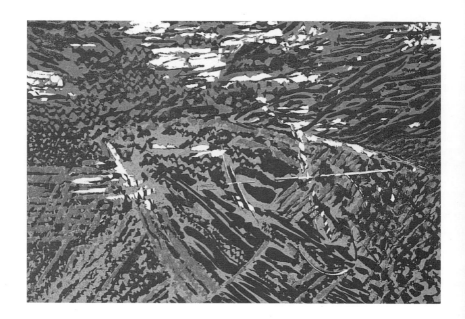

Homage to Hokusai, 2002, linocut, 32.5 x 47 cm

88 Priory Road
Hornsey
London N8 7EY

Peter Green studied at Brighton and the Institute of Education, University of London. He is a teacher-printmaker, former Head of Art Teacher Training at Hornsey College of Art and Dean of Art and Design, Middlesex University (Emeritus Professor). He is the author of books on teaching printmaking and is a former member of the Crafts Council of Great Britain. He exhibits throughout the UK, Asia and Europe. His work is in many public collections including that of the Arts Council.

Blue Land Fantasy, 2005, woodcut and stencil, 41 x 43 cm

Goose Barn
Little Casterton
Rutland PE9 4BE

email   plgreenart@waitrose.com

PHIL **GREENWOOD** RE

Phil Greenwood studied at Harrow and Hornsey Colleges of Art and has been a full-time professional printmaker since 1971. His work has been exhibited worldwide, notably in Japan. It has been shown at the Victoria and Albert Museum and in Arts Council and British Council exhibitions and is in many private and public collections including those of the Tate Gallery and other museums, universities and education authorities.

Autumn Vines, 2004, etching and aquatint – 4 colours, 48 x 59.5 cm

3 The Square
Elham
Kent CT4 6TJ

mobile   +44 (0)7885 686 670
email   saphil@ntlworld.com
web   www.philgreenwood.info

John Grigsby studied at Stoke and Leicester Colleges of Art. He taught full-time for thirty-three years, mostly in schools. Whimsical trees and interlocking field and hill shapes, with occasional bizarre figures, have for him remained constant topics. He acknowledges an ongoing debt to the work of Giorgio Morandi and Jacques Villon. He is represented in more than two dozen public collections.

Tight Squeeze, 2002, etching, 30 x 43 cm

152A Mackenzie Road
Beckenham
Kent BR3 4SD

tel.   020 8659 3951

BRIAN HANSCOMB RE

Brian Hanscomb works in the now rare medium of copperplate engraving, having served an apprenticeship in industrial printing. He works also in mixed media and pastel, having exhibited in London, Bath and abroad. His work can be viewed at New Craftsman Gallery, St Ives, and his own studio gallery in Cornwall. He has work in public collections including those of the Victoria and Albert Museum and the Science Museum, London.

Scarecrow, 1998, copperplate engraving, 29 x 19.5 cm

Tor View
Limehead
St Breward
Bodmin
Cornwall PL30 4LU

tel.   **01208 850 806**
email   **brian@brianhanscomb.co.uk**
web   www.brianhanscomb.co.uk

Marcelle Hanselaar was born in Holland and studied art at the Royal Academy of Arts in The Hague. She now lives and works in London. She works both in oil painting and etching and she shows regularly with the East West Gallery, London, and with De Queeste Kunstkamers, Watou, Belgium. She also exhibits in Holland, Germany and Australia and has won the Presse Papier Award at the Biennale Internationale d'Estampe, Quebec. Her work is in several collections including that of the Arts Council of Amsterdam and the British Museum, Prints and Drawings Collection.

Liar, Liar, house on fire, 2004, etching and aquatint, 26 x 21 cm

58 Eccleston Square
London SW1V 1PH

web   www.marcellehanselaar.com

ROGER **HARRIS** RE

Roger Harris lives and works in Gloucestershire. He studied at Kingston College of Art and specialised in printmaking at Richmond College. He exhibits widely throughout the country, his work reflecting a passion for exploring mezzotint. He also enjoys working in drypoint and wood engraving. In 1993 he was awarded 2nd prize at the Cleveland International Drawing Biennale and in 1997 gained the Pollock-Krasner Art Foundation Award (New York).

Spirit of the Sea, mezzotint, 19.5 x 14.5 cm

Springhill Cottage
Quarhouse
Brimscombe
Stroud
Gloucestershire

tel.   **01453 885426**

Brenda Hartill studied art in New Zealand and at the Central School of Art, London. Her abstract embossed etchings, collagraphs and collages are produced in her studios in London and Spain and are shown at the New Academy Gallery London, Cambridge Contemporary Art and galleries worldwide. She recently co-authored a book, *Collagraphs and Mixed Media Printmaking*, published by A&C Black.

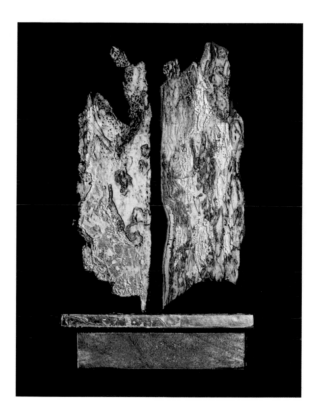

Silver Meltdown I, 1999, collaged etching, silver leaf and carborundum, 50 x 40 cm

Brenda Hartill Prints
10/11 Bishops Terrace
Kennington
London SE11 4UE

tel.   020 7582 5825 or 01424 882 942
email   brenda.hartill@gmail.com
web   www.brendahartill.com

PAUL **HAWDON** RE

Paul Hawdon studied fine art at St Martin's School of Art and undertook post-graduate study at the Royal Academy Schools. Subsequently he was awarded an Italian Government Scholarship and was also holder of the Prix de Rome in print-making. His large, often intensely-worked etchings contrast with his freer work; both reflect his love of the expressive possibilities of drawing.

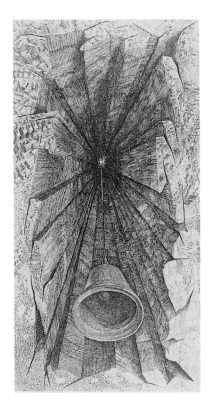

Homage to Sarah Kane, 2000, etching, 100 x 50 cm

62 Marshall Road
Cambridge CB1 7TY

tel.   01223 411 822

Jason Hicklin studied painting at St Martin's School of Art and advanced print-making at the Central School of Art in London. His work, consisting of etchings, drawings and paintings, is based on walks around the British Isles. Since 1999 he has taught etching and managed the printroom at the City and Guilds of London Art School.

Mounts Bay from Trencrom Hill Fort, 2002, etching and aquatint, 75 x 100 cm

Garden House
Meifod
Montgomeryshire SY22 6BZ
Wales

tel.   01938 500 169
email   admin@jbhicklin.freeserve.co.uk
web   www.jasonhicklin.com

STEPHEN **HOSKINS** VPRE

Stephen Hoskins is Hewlett Packard Professor of Fine Print and Director of the Centre for Fine Print Research at the University of the West of England, Bristol. He has exhibited widely throughout the world including the USA, China, Malaysia, France and Holland and his prints are based upon the structure of the kite. Hoskins has recently written his second book entitled *Inks*, published by A & C Black.

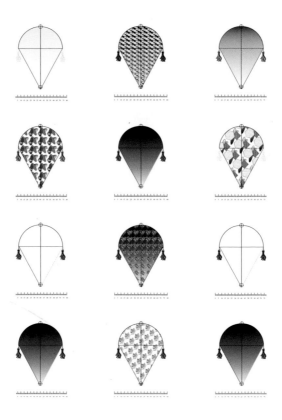

Twelve arch top kites, pigment inkjet print on Somerset enhanced paper, 101.2 x 88.5 cm

tel.   **0117 942 2497**
email   **Stephen.Hoskins@blueyonder.co.uk**
         Stephen.Hoskins@uwe.ac.uk
web   **www.uwe.ac.uk/amd/cfpr**

Edgar Holloway studied at the Slade School. His artistic talent was recognised early: his first London exhibition took place in 1931 when he was 17. Examples of his work are in many important collections, notably those of the British Museum and the Ashmolean, Oxford. He exhibits at major UK venues and regularly with Wolseley Fine Arts. *The Etchings and Engravings of Edgar Holloway* was published by Ashgate, 1996.

The Etcher, 1932, etching, 20 x 20 cm

Woodbarton
Ditchling Common
Hassocks
West Sussex
BN6 8TP

tel.   01444 244 356

JOHN HOWARD RE

John Howard worked in industry for ten years before graduating from the University of Central England with a first-class Fine Art Degree. Since then he has exhibited regularly throughout the UK. His work is held in public and private collections including the City Art Galleries of Birmingham and Manchester, the V&A, the House of Commons and St John's College, Oxford.

Trebah gardens, Cornwall, 2001, etching and aquatint, 14.8 x 17.3 cm

2 Chard Terrace
Falmouth
Cornwall TR11 2RE

tel.   **01326 314 006**
email   **johnhowardartist@hotmail.com**

Magnus Irvin studied at Hornsey College of Art and North-East London Polytechnic. He has exhibited in galleries throughout the world. As well as prints he makes films, books and sculptures and has edited a nonsense paper, *The Daily Twit*, for 25 years. Examples of his work are in the collections of the V&A, the British Museum and the Arts Council. He likes bananas, rubber gloves and Kirk Douglas.

Pin-up 2, 2005, woodcut on Okawara paper, 183 x 101.6 cm

tel.   **07906 387 239**
email   **magno@pig.abelgratis.com**

BILL JACKLIN RA HON. RE

Bill Jacklin was born in London in 1943 and relocated to New York in 1985. He graduated from the Royal College of Art in 1967 and in 1990 was elected to membership of the Royal Academy of Art. His work is represented by Marlborough Fine Art, London, and is included in collections world-wide. Recent exhibitions include *A Venetian Affair: Paintings and Monotypes*, Marlborough.

Prima della Tempesta IV, 2003, monotype, 79.6 x 59 cm

© Abe Fraindlich

^C/o Marlborough Graphics
6 Albermarle Street
London W1S 4BY

web   **www.bjacklin.com**

H J Jackson studied graphics and printmaking at Norwich School of Art under the wood engraver Geoffrey Wales. He worked in advertising for 35 years, making linocuts in his spare time. He has been a full-time printmaker since 1995, producing colour images of boats and harbours, printed by burnishing with a tobacco tin. He exhibits mainly in East Anglia, and in 2003 celebrated 50 years of linocutting.

Boats and Tractors, 2002, linocut, 42 x 45.5 cm

12 Whitehall Road
Norwich NR2 3EW

tel.   01603 611 848

JUDITH JAIDINGER RE SWE

Judith Jaidinger received her Bachelor of Fine Art degree from the School of the Art Institute of Chicago. She worked as a commercial wood engraver for Sander and Zacher Wood Engraving Company. She has participated continually in national and international exhibitions, both juried and invitational, since 1966. Her work is in collections including those of the Smithsonian Institution, Portland Art Museum and the Springfield Art Museum.

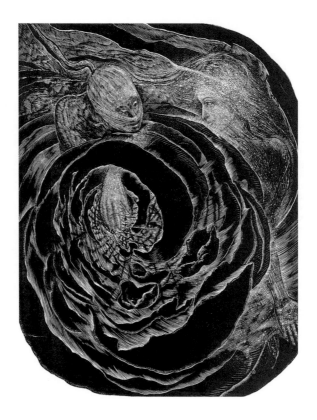

Who Can Control His Fate?, 1997, wood engraving, 28 x 23 cm

6110 N. Newburg Avenue
Chicago
IL 60631
USA

email   jaidinger@att.net

Howard Jeffs studied painting at Chelsea School of Art and completed an MA in Printmaking at Camberwell School of Art. He lectured in Printmaking and Photography at Goldsmiths College, University of London, 1972–2003. He was elected to the RE in 2002 and was awarded the Julian Trevelyan Memorial Print prize at the National Print Exhibition, London, 2003.

Black Sunflowers, 2003, monotype, 38 x 57 cm

16 Cambridge Drive
London
SE12 8AJ

Le Chenet
St Ciers 16230
Charente, France

tel.   **020 8852 2802**
mob. **07808 752 866**
email   **howardjeffs@gmail.com**

tel. + fax
0033 (0) 5 45 39 25 39

OLWEN JONES VPRWS RE RAS

Olwen Jones trained at Harrow College of Art and the Royal Academy Schools. She has taught in London Art Colleges and works freelance, her subjects reflecting her interest in complex structures, man-made and natural. Her print media are both relief and etching. Jones has regular solo exhibitions and her work is in public collections throughout the UK, including that of the National Museum of Wales.

Powerhouse at Bankside, 1995, etching, 45.7 x 38 cm

St Andrews House
Fingringhoe
Near Colchester
Essex CO5 7BG

tel.   01206 729406

Rosamund Jones studied at Leeds College of Art and produces etchings in her Yorkshire studio. She keeps sheep and cockerels and these feature prominently in her work. She has had solo exhibitions in London and Edinburgh and has shown work at the Royal Academy and the Royal Scottish Academy. She now exhibits almost exclusively at the Bankside Gallery, London. In 2000 she was awarded the St Cuthbert's Paper Mill Prize.

On my way to a party, 1980, etching, 28 x 44.5 cm

© Susan Behrens

New Bridge Farm
Birstwith
Harrogate
Yorkshire HG3 2PN

email   martingaunt@hotmail.com

STANLEY **JONES** HON. RE

Stanley Jones studied lithography at the Slade School and in Paris. He was co-founder of the Curwen Studio (1958), eventually a Director, co-founder and Trustee of its Print Study Centre and a participant in the Tate Gallery's 'Artists at Curwen' exhibition (1977). His important contribution to the making and study of prints has been recognised by the award of Southampton University's Doctorate of Letters. His work is in public and private collections.

Boreas II, 2003, monoprint, 43 x 43 cm

© Julia Hedgecoe

Lion House
6 Grange Avenue
Woodford Green
Essex IG8 9JT

Curwen Studio
Chilford Hall
Linton
Cambs. CB1 6LE

tel. **020 8504 5113**
email   info@thecurwenstudio.co.uk

tel. **01223 893 544**

Anne Jope studied painting at the Central School of Art and Design, London, and subsequently studied printmaking at postgraduate level. Besides being a Fellow of the RE she is a member of the Association of Illustrators. She has exhibited widely in Britain, notably at the Royal Academy and Art-in-Action (Art fair, Oxfordshire), and elsewhere in Europe and the USA. Her work is in various collections.

Boxwood Blocks in Bleeding Heart Yard, wood engraving, 21.2 x 16.7 cm

email   ajope@escp-eap.net

MILÁ JUDGE-FÜRSTOVÁ ARE

Milá Judge-Fürstová was born in 1975 in Czechoslovakia. She studied at Charles University in Prague and the Royal College of Art in London. She has won 14 awards for her work, including Print of the Year (Prague 2003) and the Curwen Prize (London 2004). Currently, she is the first artist in residence at the Cheltenham Ladies' College and a visiting lecturer for Coventry University. Her work, which is exhibited widely around the world, can be found in numerous private and public collections including that of Her Majesty the Queen and the V&A Museum.

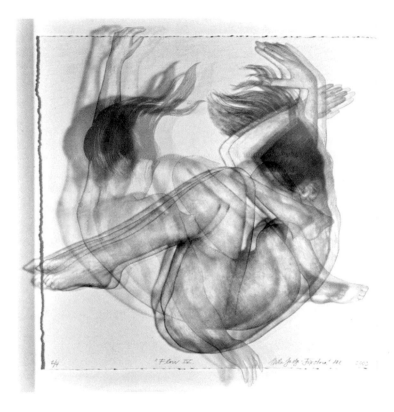

Flow IV, 2003, mixed media, 53 x 53 cm

tel.   **0775 295 8331**
email   **mila@furstova.com**
web   **www.furstova.com**

Anatoli Kalashnikov b. 1930 trained at the Moscow Institute of Industrial Arts and worked as a designer at the Likhachev Automobile Plant. Whilst art editor at the Soviet Writer Publishing Company he began to get bookplate commissions and eventually his clientèle – and his exhibiting – became international. He heads the Bookplates and Book Arts section of the International Commonwealth of Bibliophiles.

Untitled, wood engraving

C/o Geri Waddington
3 West Street
Oundle
Northants PE8 4EJ

email   g.waddington@dial.pipex.com

Paul Kershaw trained for the teaching profession in Newcastle upon Tyne and spent twelve years as an art teacher before leaving to concentrate on his own print-making. He began wood engraving, self taught, in 1985, became a member of the Society of Wood Engravers in 1990 and exhibits throughout the UK. He runs the Orbost Gallery on the Isle of Skye.

Sheer Delight – Stac Lee, 2001, wood engraving, 13.4 x 9.7 cm

Altvaid
Harlosh
Isle of Skye IV55 8ZF
Scotland

tel.   **01470 521 288**
web   **www.plkershaw.co.uk**

Ken Kiff taught part-time in secondary schools and subsequently at Chelsea and the Royal College of Art. Almost annual one-man exhibitions from 1979 until his death in 2001 were mounted in many major British centres as well as in New York and in 1992–4 he was Associate Artist at the National Gallery, London. He also participated – again almost annually for thirty years - in important mixed exhibitions throughout Britain, Europe and America.

Walking Past Rock & Flowers, 1995, etching, 31 x 35 cm

© Toby Newman

c/o Marlborough Graphics
6 Albemarle Street
London W1S 4BY

tel.   020 7629 5161
web   www.marlboroughfineart.com

ROBERT **KIPNISS** RE NA

Robert Kipniss has had over 180 one-man shows in such cities as London, Berlin, Tokyo, Lima, New York and San Francisco. His work is in the collections of the British Museum, the French Bibliothèque Nationale, the New York Metropolitan Museum and various other national institutions world-wide. A retrospective exhibition will take place in New Orleans in 2006 and will subsequently tour other American museums.

Vase, Chair & Trees, 1995, mezzotint, 24 x 17.7 cm

PO Box 112
Ardsley-on-Hudson
New York 10503
USA

email   rkipniss@msn.com

Anita Klein studied at Chelsea and the Slade Schools of Art. She is President of the Royal Society of Painter-Printmakers and shows paintings and prints regularly with the Boundary Gallery and Advanced Graphics London. She also exhibits in galleries throughout the UK, Europe, the US and Australia. Her work is in many public collections, including that of the Arts Council England.

Nige isn't tired, 2004, drypoint, 30 x 40 cm

© Nigel Swift

tel.   020 8691 2374
fax   020 8694 9185
email   anita@anitaklein.com
web   www.anitaklein.com

LINDA **LANDERS** RE

Linda Landers is a wood engraver and maker of limited edition artist's books. She is also an essayist and poet and has been awarded prizes both for her poetry and her printmaking. Her hand-produced books are in private and public collections in Britain and abroad including those of the V&A, the Ashmolean, the British Library and the Museum van het Boek (Netherlands). Tate Britain and the Royal Academy have shown her work.

Boy with Owl, 2004, linocut, 86 x 76 cm

mobile   **07958 608 856**
email   lindalanders87@hotmail.com

Karolina Larusdottir is originally from Iceland. She studied at the Ruskin School of Art, Oxford, from 1965 to 1967. She is a member of the New English Art Club. Her awards include that of the Artist of the Year 2005, Islensk Grafik. She has exhibited at the Pringle Gallery, Philadelphia; the Castle Gallery, Inverness; and the Royal Academy Summer Exhibitions. Solo shows have been held at Cambridge Contemporary Art and the Sarah Wiseman Gallery, Oxford.

The Big Stage, 2004, etching and aquatint, 40 x 50 cm

tel.   **01223 369 269**
email   **karolina.larusdottir@ntworld.com**
web   **www.karolinalarusdottir.com**

JOHN **LAWRENCE** RE SWE

John Lawrence, book illustrator and wood engraver, studied at Hastings School of Art and London's Central School. He has illustrated more than 150 books and his work is represented in the V&A, the Ashmolean, Manchester Metropolitan University and collections in the USA. He taught illustration at Camberwell School of Art, was Visiting Professor at the London Institute and is at present teaching an MA course on Children's Books at APU Cambridge School of Art.

Once Upon a Time, 1990, wood engraving, 20.4 x 15.2 cm

6 Worts Causeway
Cambridge CB1 8RL

tel.+fax   **01223 247 449**

Peter Lawrence was born in 1951. He gained a degree in Graphic Design (Illustration) from Bristol and has worked as a freelance illustrator and book designer since moving to Oxford in 1978. He began wood engraving in 1990 and was elected a SWE member in 1999. He has won awards at the National Print Exhibition, London, and the SWE's Rachel Reckitt Prize for engraving. He is currently Managing Director of Oxford Designers & Illustrators.

Creative Review, 2003, wood engraving, 15 x 26 cm

48 Lonsdale Road
Oxford OX2 7EP

tel.   **01865 559 193 (Home)**
tel.   **01865 512 331 (Work)**
email   **pete@odi-design.co.uk**

URSULA **LEACH** RE

Ursula Leach studied at Winchester, Wimbledon and Farnham Schools of Art. She shows paintings and prints with the Attendi Gallery and Oliver Contemporary, London, as well as with other galleries throughout Britain. Her work is in many hospitals and other public collections and it focuses on landscape and current agricultural practices. Firstly, though, her concern is with the formal issues of image-making.

Hedges/Edges, 2004, carborundum and hand colouring, 36 x 36 cm

14 The Square
Cranborne
Dorset BH21 5PR

email   ursulaleach@hotmail.com
web   www.ursulaleach.com

Ann Le Bas has always worked freelance and has undertaken various commissions for clients including the National Trust Foundation for Art, the Bishop of Bath and Wells, Royal Academy Graphics, etc. A collection of her prints is included in the permanent collection of the Ashmolean Museum (Dept. of Western Art) and she is represented in the Fitzwilliam Musem.

San Giorgio & the Lagoon Sunrise, 1992, aquatint, 23.5 x 49 cm

Winsford
Nr Minehead
Somerset
TA24 7JE

tel.   **01643 851 217**
email   **annlebas@winsfordcentre.org.uk**

JEAN **LODGE** RE

Jean Lodge was born in Ohio in 1941. She moved to Europe in 1963 after completing a degree in literature. In England, she studied at the Ruskin School of Drawing and then worked with S W Hayter for three years in Paris. She has had her own printmaking atelier in Paris since 1969 and was Head of Printmaking at the Ruskin School from 1978-1996. She continues to divide her time between England and France and exhibits widely throughout the world.

L'image du promeneur sur l'eau, 2004, woodcut, 50 x 65 cm

© Michael Gabriel

52 Granville Court	18 Rue Ernest
Cheney Lane	Cresson 75014
Headington	Paris
Oxford OX3 0HS	France

Irvine Loudon was trained in printmaking at the Oxford Printmakers' Cooperative, of which he is a member. He is also a member of the Oxford Art Society. He exhibits regularly in exhibitions held by these Societies and by the RE and has held one-person exhibitions at various places including in London, Oxford, Manchester, Newbury and Marlborough. He is also an historian of medicine.

Auvergne II, 1992, etching, 24.5 x 23 cm

The Mill House
Locks Lane
Wantage
Oxford OX12 9EH

tel.   01235 763 965
fax   01235 766 696
email   Irvine.loudon@wuhmo.ox.ac.uk

ROBIN **MACFARLAN** ARE

Robin MacFarlan studied at Somerset and Norwich Schools of Art. He subsequently worked as an illustrator for clients such as *The Guardian* and *The Financial Times* whilst developing his own ideas as a printmaker. His first solo exhibition was held in 1973 since when his work has been exhibited widely, both in Britain and abroad.

They Met Near London Zoo, 2002, etching, 18 x 38 cm

c/o Bankside Gallery
48 Hopton Street
London SE1 9JH

tel.   020 7928 7521
mob.   07815 473 179
email   robin.macfarlan@btinternet.com

LEONARD McCOMB RA RWS RP HON. RE

Leonard McComb studied in Manchester and at the Slade School. He was Keeper of the Royal Academy Schools 1995-1998. His work has been exhibited at the Hayward Gallery, the Oxford Museum of Modern Art, the Serpentine Gallery, the Whitechapel Gallery and the Venice Biennale. He is represented in many private and public collections including those of the British Museum, the V&A and the Tate Gallery, and in 1997 was awarded the Nordstern Printmaking Prize.

Bronze Horse, St Mark's Venice, 1990, etching, 39 x 47 cm

4 Blenheim Studios
29 Blenheim Gardens
London SW2 5EU

tel.   020 8671 5510

SALLY McLAREN RE

Sally McLaren studied at the Ruskin School of Art, Oxford, and the Central School of Art, London, gaining a French Government Scholarship to work at Atelier 17 in Paris with S W Hayter. She has exhibited world-wide. Her work is in many public collections including those of the New York Public Library, the Scottish Arts Council, the Ashmolean Museum, Oxford, and the Government Art Collection.

Elements of Oceans, etching and chine collé, 74.9 x 106.6 cm

Steeple Close
Milton
East Knoyle
Nr Salisbury
Wilts SP3 6BG

tel.   **01747 830 495**
email   sally@sallymclaren.co.uk
web   www.sallymclaren.co.uk

Tiffany McNab studied at the Royal Melbourne Institute of Technology, completing a Bachelor of Arts Degree in Painting and a Diploma in Printmaking. Her work has been shown in many galleries and major competitive exhibitions throughout the UK and Australia. McNab lives in Melbourne where she works from her own studio, producing etchings and watercolours that reflect her love of still life.

Darwin's Beetle, 2004, 3-plate colour etching with silver leaf, 49 x 42 cm

8 Gellibrand Street
Kew
Victoria
Australia 3101

tel.   0061 3 9853 6181
email   mcnabt@bigpond.net.au

JOHN A **MCPAKE** RE PMAFA

John McPake is currently President of the Manchester Academy of Fine Arts. He works predominantly in drawing, etching and acrylic painting and has been a member of the RE since 1978. He has exhibited widely and has executed a number of commissions. He has work in public and private collections in the UK, France, USA, Germany, Norway, Holland, Australia, South Africa and Russia.

Dark Fable, 2002, etching, 30 x 25 cm

21 Ingbirchworth Road
Thurlstone
near Sheffield S36 9QN

tel.   **01226 765 192**
email   johnmcpake@beeb.net

Sasa Marinkov studied Fine Art at Leeds University and Printmaking at the Central School of Art and Design. She has exhibited her prints regularly in the UK and abroad, including several international print biennales. She has won six major awards for her relief prints and has work in public collections including the Arts Council Collection, UK.

At the Maeght, 2002, woodcut, 57.5 x 88 cm

"Woodcut"
Riverside
Twickenham
TW1 3DJ

tel.   **020 8891 2175**
email   sasamarinkov@woodcut.free-online.co.uk
web   **www.axisartists.org**

TONI **MARTINA** RE

Toni Martina is a Prix de Rome Scholar and studied at Kingston and the Central School of Art and Design, London. He has exhibited in the UK and internationally. His work is in public collections including those of the New York Public Library and the Museum of London. Martina won the Birgit Skiöld prize recently, adding to a long list of achievements.

Bar Italia, Soho, etching, 46 x 61 cm

tel.   **01424 434 983**
mob.   **07969 860 607**
email   **t.martina@btinternet.com**

Mike Middleton studied at Farnham Art School, Sheffield Polytechnic and Chelsea School of Art. He was elected to the RE in 1981. Currently he is Head of Printmaking at Colchester School of Art. He exhibits regularly in the UK and his work is in public and private collections.

Sculpture Shadows, 2004, digital print, 37 x 28 cm

Gable End
39 Peldon Road
Abberton
Colchester, Essex CO5 7PB

tel.   01206 736 140
email   mike.middleton@btinternet.com

JULIA **MIDGLEY** RE

Julia Midgley is Reader in Documentary Drawing at Liverpool School of Art and Design, John Moores University. Her etchings, drawings and paintings are exhibited widely in the UK and abroad. Whilst recently Artist in Residence for the Excavation of Chester Roman Amphitheatre, Midgley's work features in public and private collections including those of the Wellcome Foundation and the National Museum of Science and Industry.

Hindsight, 2003, etching and aquatint, 25 x 30 cm

tel.   **01606 77006**
email   **julia@juliamidgley.co.uk**
web   **www.juliamidgley.co.uk**

Stephen Mumberson studied at Brighton University and the Royal College of Art. He is Reader in Fine Art Printmaking at Middlesex University, Deputy Chairman of the Printmakers' Council and a writer on printmaking. He has worked in Europe and Africa as a printmaker and exhibits internationally. His interests span etching, stone lithography, digital print and relief printmaking.

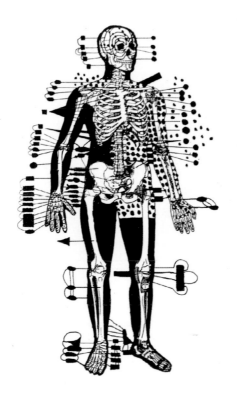

New Body, 2003, digital print, 35.5 x 22.8 cm

Flat C
18 Towpath Walk
London E9 5HX

email   s.mumberson@mdx.ac.uk

BRENDAN **NEILAND** HON. RE

Brendan Neiland studied at Birmingham and the Royal College of Art; he was awarded the RCA's Silver Medal. His work has been handled by a number of London galleries, latterly the Redfern. He has exhibited internationally and examples of his work are held in many major institutions. In 1996 he was appointed Professor of Painting, University of Brighton, and was from 1998-2004 Keeper of the Royal Academy of Art.

Beijing Garden, 2003-4, screenprint, 76 x 53 cm

Studio tel. **020 8673 4597**
Redfern Gallery tel. **020 7734 1732**
Studio, France tel. **0033 5 46 01 62 97**
web **www.brendanneiland.com**

Barbara Newcomb received a degree with distinction in painting from Syracuse University, New York, and a diploma in etching with distinction at the Central School, London. She also studied with S W Hayter in Paris on a full scholarship. Many institutions have collected her work, among them the Arnolfini, Vienna; the Victoria and Albert Museum, London; the Arts Council of Great Britain; and the Bibliothèque Nationale, Paris.

Percussion, 2004, monoprint, 87.6 x 83.8 cm

12 Milman Road
Reading RG2 0AY

tel.   0118 954 9925

JACKIE **NEWELL** RE

Jackie Newell gained a first degree in Fine Art from New York State University and an MA from Camberwell School of Art. She has exhibited extensively, internationally, during the last 15 years and her work is in collections both private and public in Britain and the USA. Since returning to the UK she has worked as an art lecturer in further and higher education and is co-author of a book entitled *Monoprinting*, published by A&C Black. She currently teaches in two of Her Majesty's prisons.

Urban Aspects, 2002, etching and aquatint, 15.3 x 11.5 cm

email   jackie1_newell@onetel.com

Sarah van Niekerk studied at the Central and Slade Schools of Art. She exhibits widely in Britain and abroad, twenty of her exhibitions having been solo, and her work is in public collections in Britain (notably the V&A), Russia, America and Canada. She has taught at the Royal Academy Schools and elsewhere and a monograph of her work was published by the Primrose Academy Press in 2000.

Willows from Charlbury Station, 1981, wood engraving, 12.7 x 10 cm

Priding House
Saul
Gloucestershire GL2 7LG

tel.+fax   01452 740 285

KENNETH H **OLIVER** RE RWS RWA

Kenneth Oliver studied at Norwich School of Art and the Royal College of Art. He was head of printmaking studies at the Gloucestershire College of Art and Design, Cheltenham. He is a member of the Royal Society of Painter-Printmakers, the Royal Watercolour Society and the Royal West of England Academy, and exhibits in UK galleries, with work in public and private collections.

Winter on Cleeve Hill, Cheltenham, 1960, etching and aquatint, 35.5 x 25.4 cm

Green Banks
Blakewell Mead
Painswick
Stroud
Gloucestershire GL6 6UR

Chris Orr is currently Professor of Printmaking at the Royal College of Art and exhibits his paintings, drawings and prints throughout the world. Besides at the Royal Academy, he shows in London with the Jill George Gallery. His work is in the collections of the British Museum, Royal Academy of Arts, Victoria and Albert Museum, Science Museum and many private collections.

I Want To Dress Up In Greengrocer Clothes, 2001, lithograph and screenprint, 65 x 52 cm

5 Anhalt Road
Battersea
London SW11 4NZ

tel.   020 7738 1203
email   chrisorr@aol.com

ANA MARIA **PACHECO** RE

Ana Maria Pacheco (b. Brazil 1943) first came to London in 1973 on a British Council Scholarship to the Slade School; since then she has lived and worked in England. From 1997–2000 she was Associate Artist at the National Gallery, London. Her work has been exhibited widely in the UK and abroad. Recent venues include: National Gallery, London; British Museum; Ashmolean Museum, Oxford; Kunsthalle Wien, Austria; Salander-O'Reilly Galleries, New York.

Domestic Scenes 8, 2000, etching with drypoint, 20 x 26.8 cm

© Bartolomeu dos Santos

c/o Pratt Contemporary Art
The Gallery
Ightham
Sevenoaks
Kent TN15 9HH

tel.   **01732 882326**
email   pca@prattcontemporaryart.co.uk
web   www.anamariapacheco.co.uk

Celia Paul, born in India, studied at the Slade School, London. Marlborough Graphics and Marlborough Fine Art have held solo exhibitions of her etchings, drawings and watercolours and she has been represented in important group shows in London, Paris, Vienna, Madrid and Granada. Her work has been widely reviewed in such journals as *The Art Newspaper*, *Time Out*, *The Observer* and *The Financial Times*.

My Mother Seated, 1997, softground etching, 29.5 x 23.5 cm

c/o Marlborough Graphics
6 Albemarle Street
London W1S 4BY

tel.   020 7629 5161
web   www.marlboroughfineart.com

Hilary Paynter's work ranges extensively from dramatic landscapes and strange geological forms to perceptive socio-political comment which may be subtle or savage. In between are the gentler, reflective waterscapes, castles and other images. A recent commission features a suite of her wood engravings enlarged to 2 x 22 metres and printed on vitreous enamel panels installed in Newcastle's Central Station.

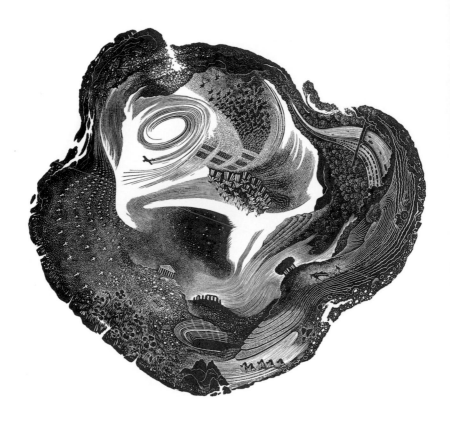

Tree with a Long Memory, 2003, wood engraving, 17 x 18 cm

Torridge House
Torridge Hill
Bideford
Devon EX39 2AZ

tel.+ fax **01237 479 679**
email   **hilarypaynter@yahoo.com**

Melvyn Petterson studied fine art and postgraduate printmaking at Camberwell School of Art. In 1992 he established, with Colin Gale, the Artichoke Workshop in Brixton. Of his books on printmaking, *The Instant Printmaker* (Collins & Brown) is the most recent. His work has gained many awards, has been exhibited widely in Britain and abroad and examples are held by many major galleries and museums including Oxford's Ashmolean.

After the Storm, Richmond Park, 1996, etching, 61 x 91.4 cm

Artichoke Print Workshop
Unit 51
Shakespeare Business Centre
245a Coldharbour Lane
London SW9 8RR

email   mpetterson@btinternet.com

NEIL **PITTAWAY** RWS RE

Neil **Pittaway** completed his studies at the Royal Academy Schools, Bradford University and Gloucester University. His work has been exhibited in Britain, Italy and India and is in the collection of the British Museum, the Ashmolean, the V&A, the Guildhall, the Dover Street Arts Club, St Paul's Cathedral and other prominent institutions. His work features in *London in Poetry and Prose* (Enitharmon Press, 2003) and 'Great Artists' (Channel 5 TV, 2001).

Cutty Sark, 2000, etching, 60 x 70 cm

1 Glenfields
Netherton
Wakefield
West Yorkshire
WF4 4SH

tel. 01924 274 502
mob. 0770 841 8881
email njpittaway@hotmail.com

Bryan Poole studied botanical art at Kew Gardens, London. He works in a 16th-century technique after the French master Redouté, inking his plates in multiple colours to produce a fully-chromatic image at a single printing. He is a regular exhibitor at the Chelsea Flower Show and has work in the Smithsonian Institution, Washington; the Hunt Institute, Pittsburgh; the Natural History Museum, London; and Kew Gardens.

Banana Palm - Flower and Fruit, 2002, aquatint etching, 70.5 x 37.5 cm

tel.	**020 7254 1213**
fax	**020 7923 1616**
email	**bryan@etchart.co.uk**
website	**www.etchart.co.uk**

TREVOR **PRICE** RE

Trevor Price studied at Winchester and Falmouth Schools of Art. He exhibits widely throughout Britain, Europe, New Zealand and Australia and regularly has solo shows of paintings as well as prints. His work is in many collections including those of the Bank of England and Yale University. He has studios in London and St Ives, Cornwall, and shares his time between the two.

Comfortable Silence II, 2004, drypoint, 39 x 43 cm

email   trevor@trevorprice.co.uk
web   www.trevorprice.co.uk

SIMON **REDINGTON** RE

Simon Redington studied Fine Art at Goldsmiths and Central St Martin's College, London. A subterranean anthropologist by nature, his imagery generally contains scenes of lurid, urban realities. In 1993 he founded the Kamikaze Press to produce folios of work using letterpress, woodcuts and etchings. He now lives in Hanoi, Vietnam, where his outlook has developed a broader world vision.

Tiger, 2004, woodcut, 110 x 80 cm

email   simonredington@hotmail.com
web   www.kamikazepress.com

PHILIP **REEVES** RSA PPRSW RGI RE

Philip Reeves studied at Cheltenham School of Art and the Royal College of Art. He founded the Edinburgh Printmakers' Workshop in 1967 and Glasgow Print Studio in 1972. Works in public and private collections include: British Government Art Collection; Contemporary Art Society; Hunterian Art Gallery, Glasgow; Scottish National Gallery of Modern Art; Fleming–Wyfold Art Foundation; Victoria and Albert Museum; Scottish Arts Council.

Twin Stacks, 2003, etching and aquatint, 33 x 68.5 cm

© Peter Black

13 Hamilton Drive
Glasgow G12 8DN

tel.   0141 339 0720

Paula Rego, born in Lisbon, studied at the Slade School, London. In 1990 she became the National Gallery's first Associate Artist. Her wide reputation has gained her three Honorary Doctorates from, respectively, the Universities of St. Andrew's and East Anglia and from Rhode Island School of Design. She has had solo exhibitions in Britain, Portugal, Spain, Australia and North America, participated in very many mixed shows and been the subject of several monographs and broadcasts. *Paula Rego: The Complete Graphic Works* by T G Rosenthal (Thames & Hudson, London, 2003) has recently been published in paperback.

Three Blind Mice, from *Nursery Rhymes*, 1989, etching and aquatint, 52 x 38 cm

c/o Marlborough Graphics
6 Albemarle Street
London W1S 4BY

tel.   020 7629 5161
web   www.marlboroughfineart.com

RAY **RICHARDSON** ARE

Ray Richardson was born in London in 1964. He studied at Goldsmiths College. He has made prints with professional studios regularly since 1991, specifically etchings and lithographs with Glasgow Print Studio and silkscreens and monoprints with Advanced Graphics London. Currently based in London, Richardson has lived and exhibited in Brussels and shows frequently in Paris, Los Angeles and Tokyo.

A London Film, 2002, screenprint, 46 x 45.5 cm

c/o Advanced Graphics London
32 Long Lane
London SE1 4AY

tel.   020 7407 2055
web   www.advancedgraphics.co.uk

Bob Saich joined Advanced Graphics London in 1971, after working for three years at Kelpra Studios. Currently a partner there, he has worked with artists such as Neil Canning, Anthony Frost, Albert Irvin, Anita Klein and Ray Richardson. He is well known as a master printmaker, pioneering screenprint and woodblock techniques for artists both commissioned by and published by Advanced Graphics London.

Borough II, 2004, screenprint, 52 x 43.5 cm, print by Albert Irvin

c/o Advanced Graphics London
32 Long Lane
London SE1 4AY

tel.   020 7407 2055
web   www.advancedgraphics.co.uk

BARTOLOMEU **DOS SANTOS** RE

Bartolomeu dos Santos studied at the Slade School with Anthony Gross RE, whom he succeeded as Professor of Printmaking (1961–1996). He is now Emeritus Professor. His work is in the British Museum, the V&A, Museum of Modern Art (New York), Bibliothèque Nationale (Paris), Gulbenkian Foundation (Lisbon) and other collections. It has also been shown in solo and mixed exhibitions world-wide. Public art includes large etched stone decorations in Lisbon, Macao and Tokyo.

Ne Pas Effacer I, 2000, photo etching, 70 x 101 cm

© J. Cutileiro

57 Talbot Road
London N6 4QX

tel.   020 8348 0416
fax   020 8340 2721
mobile 00351 962 409 066

Nana Shiomi came to London in 1989 to study at the Royal College of Art, after having practised printmaking for several years in her native Japan. Her theme is 'Art about Culture' or 'Art about Art'. After re-examining Western culture in the early 1990s she has now moved on to a consideration of her own Japanese culture.

A Room on the Other Shore - Moon, 2003, woodcut, 61 x 90 cm

96A Greenvale Road
Eltham
London SE9 1PF

web   www.nanashiomi.com

SHEILA **SLOSS** ARE

Sheila Sloss was trained at Chelsea School of Art (Post Graduate) and the Royal College of Art. An experienced approach to printmaking and wide technical knowledge enable Sloss to mix photo etching, collage, screenprint and her own photographs. She works at the London College of Communication, formerly the London College of Printing, and also exhibits with the Printmakers' Council.

Dolls House Revisited, 1999, photo etching with chine collé, 55 x 50 cm

37 Lawrence Road
London W5 4XJ

email   s.sloss@lcc.arts.ac.uk

Ian Stephens studied illustration at Northampton School of Art. He became an RE fellow and Society of Wood Engravers member in 1984. He exhibits widely and many private and public collections in Britain and overseas (including the Ashmolean Museum, Oxford) hold examples of his work. He works as a freelance illustrator (regularly for the Folio Society) and printmaker.

Saling, 1997, wood engraving, 10 x 8.8 cm

46 Yardley Drive
Northampton NN2 8PE

tel.   01604 460 457

JANE **STOBART**

Jane Stobart is an artist-printmaker. She teaches at Goldsmiths College and she regularly exhibits in the UK and abroad. Her work appears in private and public collections around the world, including those of the Smithsonian Institute, the London Museum, and the Fitzwilliam and Ashmolean Museums. Stobart is the author of *Printmaking for Beginners* and *Drawing Matters*, A&C Black.

Two Bells, 2004, aquatint, 36.5 x 29 cm

47 Potter Street
Harlow
Essex
CM17 9AE

email   jstobart.printmaker@virgin.net

Andrew Stock is a self-taught painter and printmaker. He is the President of the Society of Wildlife Artists and shows work regularly at the Mall Galleries, the Tryon Gallery and his studio in Dorset. He also exhibits in galleries throughout the UK and abroad. He has undertaken many commissions and book illustrations and has work in several private collections including that of the Sultan of Oman.

Eglise de St Pierre, 2002, etching and aquatint, 25.4 x 20.3 cm

The Old Schoolhouse
Ryme Intrinseca
Sherborne
Dorset DT9 6JX

tel.   **01935 873 620**
email   AndrewNstock@aol.com
web   www.andrewstock.co.uk

EVA **STOCKHAUS** HON. RE

Eva Stockhaus studied art in her native Sweden. In 1985 she joined the Society of Wood Engravers. She is also a member of Grafiska Sällskapet, the Swedish Print Society, and of the Swedish section of the International Xylon. Nature is the chief inspiration for her work, which is represented in the Swedish National Museum, New York Public Library, Vienna's Albertina and the Victoria and Albert Museum.

Late Summer, wood engraving

Pontonjärgatan 16
112 37 Stockholm
Sweden

Sandy Sykes studied as a painter–printmaker and maker of books. She won residencies in New York (2000), Japan (2001), a Commissions East Award (2002) and in 2003 a major Arts Council England East Award and the Annual Lorne Award Scholarship. She exhibits widely both nationally and internationally. Her work is in many collections including those of Tate Britain and The Museum of Modern Art New York.

Tales for Innocents, 2000, mixed media, 76 x 57 cm

email   sandy@sandysykes.co.uk
web   www.sandysykes.co.uk

PAUL **THIRKELL** PHD ARE

Paul Thirkell studied and practised as a printmaker in Sydney, Australia, until 1994. After spending a year working at Chelsea College of Art he linked up with the University of the West of England School of Art, Media and Design, where he undertook one of the first print-based doctorates in Britain. Since gaining this in 2000 he has been an integral member of UWE's Centre for Fine Print Research and currently holds the position of Senior Research Fellow. Thirkell's specialist areas of practice include collotype and wide-format digital printing.

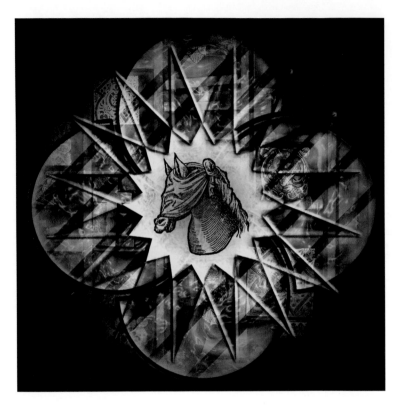

A Nod is as Good as a Wink, 2004, pigmented inkjet, 9 x 9 cm

27 Sydenham Road
Cotham BS6 5SJ
Bristol

tel.   0117 924 1565
email   Paul.Thirkell@uwe.ac.uk

Glynn Thomas studied at Cambridge School of Art and on graduating became Head of Printmaking at Ipswich School of Art. Since 1979 he has been a full-time professional artist. His work is in the Window Gallery, Brighton, and the CCA, Cambridge. He has exhibited at London's Barbican Centre, Manchester Royal Exchange and Leeds Craft Centre. Work has been commissioned by Christie's Contemporary Art and is in a number of public collections.

Burano, 2002, etching, 26 x 14 cm

email   studio@glynnthomas.com
web   www.glynnthomas.com

JAMES **TODD** RE SWE

James Todd is Professor Emeritus of Art and Humanities at the University of Montana. He has exhibited his work throughout North and South America, England, Europe and Asia, and is represented in many public collections. Throughout his career he has devoted his art to political and social themes, and his writing to topics in the social history of art.

Pool Sharks, 2004, hand coloured woodcut, 45.7 x 60.9 cm

6917 Siesta Drive
Missoula
Montana
79802 USA

tel.   406 543 7045
email   jktodd@mcps.k12.mt.us

Ann Tout studied at the Bournemouth Municipal College of Art. From 2000 she has been the Society of Wood Engravers' Honorary Secretary and in 2003 she won one of the Society's awards in the National Print Exhibition, London. Tout is also renowned as a bookbinder and watercolourist and has recently completed a large oil painting in a series of church commissions.

Paradise Retained, 2002, wood engraving, 17.6 x 10.1 cm

33 Union Street
Trowbridge
Wiltshire BA14 8RY

email   ann@anntout.com

GEORGE W **TUTE** RE RWA

George Tute studied at Blackpool School of Art and the Royal Academy Schools, 1950–58. He taught at York School of Art and the University of the West of England and is a member of the Royal West of England Academy (on whose website his work may be seen). A painter, illustrator and printmaker, he produces wood engravings for private and commercial clients. He exhibits in Bristol and London.

When earth's last picture is painted, 2004, wood engraving, 10 x 10 cm

46 Eastfield
Bristol BS9 4BE

email   georgetute@onetel.net
web   www.georgetute.com

Ruth Uglow studied BA (Hons) Printmaking at Winchester School of Art and Design and MA Printmaking at the Royal College of Art. She was elected to the Royal Society of Painter-Printmakers in 2001. She has won numerous awards including the Royal Overseas League Travel Scholarship (2003), the Printmaking Today Prize (2002) and the Desmond Preston Drawing Prize (2000). Her work is exhibited in galleries throughout the UK.

Claustrophobia, 2001, etching, aquatint and drypoint, 60 x 100 cm

tel.   07799 101 589
email   ruglow@hotmail.com

BREN **UNWIN** ARE

Bren Unwin received the Gwen May student award from the RE in 2004. She has an MA (Research) in Fine Art Practice and is currently engaged in PhD research at the University of Hertfordshire. Unwin is a member of the Newlyn Society of Artists and the Penwith Society of Arts. Examples of her work are held in British and international collections.

Point of View, 2004, etching, 62 x 38 cm

tel.   07785 960 291
email   b.unwin@herts.ac.uk

Carol Walklin studied graphic design at the Royal College of Art, London. Her association with the RE has been long, with participation in members' shows and thematic exhibitions. Her prints are in private collections worldwide. She continues working at her linocuts, woodcuts and etchings, always pursuing excellence through the expression of her very individual ideas and colourful imagery.

Cross Cochin and family, 2000, linocut, 47 x 38 cm

2 Thornton Dene
Beckenham
Kent BR3 3ND

tel.   020 8658 2086

NORMAN **WEBSTER** RWS RE ARCA

Norman Webster studied at Tunbridge Wells Art School and the Royal College of Art. He taught at Leeds College of Art and then at the Leeds Polytechnic where he retired as Senior Lecturer in Printmaking. He has shown etchings extensively over the years in various exhibitions in the UK and abroad. Many of his prints are in public and permanent collections.

Ripon Cathedral from the East, etching, 34.5 x 46.2 cm

48 The Drive
Cross Gates
Leeds
West Yorkshire LS15 8EP

Frans Wesselman studied art-teaching, printmaking and photography in the Netherlands. He now lives and works in England and makes etchings, paintings and stained glass. His work is usually narrative in nature and examples are held in many private and public collections both in Britain and abroad. He is a member of the Worcester Guild of Designer-Craftsmen and a member of the Royal Birmingham Society of Artists.

Letter, 2002, etching with woodcut, 30.4 x 30.4 cm

5 Green Mount
Cunnery Road
Church Stretton SY6 6AQ

email   frans@wesselman.freeserve.co.uk
web   www.fwstainedglass.com

JIM **WESTERGARD** RE SWE

Jim Westergard has been creating prints and drawings since graduating from university. In 1969 he began teaching at universities and colleges in Colorado and Illinois. He moved to Canada in 1975 and continued to teach there. He is now retired from teaching but continues to produce wood engravings in his own studio. He has been a regular international exhibitor throughout his career.

The Artist as an American, 1962, wood engraving, 20.5 x 15.2 cm

14 Munro Crescent
Red Deer
Alberta T4N 0JI
Canada

tel.   403 343 1772
email   jimwest@telusplanet.net
web   www.telusplanet.net/public/jimwest

Judy Willoughby studied fine art at Birmingham, Leicester and Bristol Art Schools. She uses a variety of media to produce colourful semi-abstract images in her studio in Somerset. Her work is exhibited widely and is included in public collections as well as in schools, hospitals and hotels.

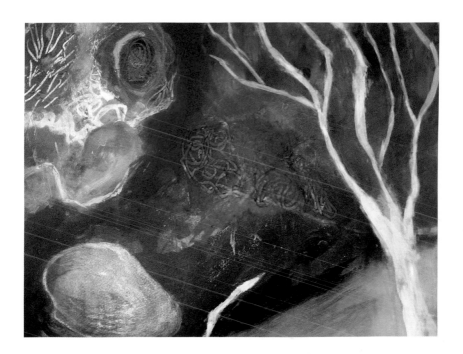

Rock Garden, 2004, monoprint and screenprint, 32 x 38 cm

The Old School
Sand Street
Milverton, Taunton
Somerset TA4 1JN
tel.   **01823 401 130**
email   willoughbyjudy@hotmail.com
web   www.judywilloughby.co.uk
        www.judywilloughby.com

JOSEPH **WINKELMAN** PPRE HON. RWS RWA

Joseph Winkelman came to Oxford University from America in 1968 to train at the Ruskin School of Drawing and has been a printmaker here ever since. He served as President of the RE 1989-1995 and is on the Board of Directors of Bankside Gallery. He exhibits widely, winning prizes internationally for miniature etchings, and is represented in many notable public collections.

Liatach, 1999, aquatint, 30 x 45 cm

The Hermitage
69 Old High Street
Headington
Oxford OX3 9HT

tel.   **01865 762 839**
mobile **07905 759 635**
email   **winkelman@ukgateway.net**

P. K, Burt & Co
57 Union Street
London
SE1 0AS
tel. 020 7407 6474
www.rkburt.co.uk
Fine art paper merchant

John Purcell Paper
15 Rumsey Road
London
SW9 0TR
tel. 020 7737 5199
www.johnpurcell.net
Fine art paper merchant, supplier of T.W.
screenprinting inks

Intaglio Printmaker
62 Southwark Bridge Road
London
SE1 0AS
tel. 020 7928 2633
www.intaglioprintmaker.com
General printmaking suppliers, inks, rollers,
sundries etc

T.N. Lawrence
Mail order and Hove shop
208 Portland Road
Hove
Sussex
BN3 5QT
tel. 01273 260260
www.lawrence.co.uk
General printmaking suppliers, inks, rollers,
sundries etc

Chris Daunt
1 Monkridge Gardens
Dunston
Gateshead
NE11 9XE
tel. 0191 420 8975
www.chrisdauntwoodengravingblocks.co.uk
Supplier of wood engraving blocks and bespoke
tools. Block resurfacing and tool sharpening
services.

Cornelissen
105 Great Russell Street
London
WC1V 3RY
tel. 020 7636 1045
www.cornelissen.com
Tools, inks, quality pigments and copper
plate oils

Openshaw Ltd
Woodhouse Road
Todmorden
OL14 5TP
tel. 01706 811421
www.openshaw.com
Suppliers of chemicals and acids, Jet plate
zinc and pre-sensitized photo etching zinc
and copper

Richards of Hull
Unit 1
Acorn Estate
Bontoft Avenue
Hull
HU5 4HF
tel. 01402 442422
www.richards.co.uk
Tailor made acid baths, sinks, ventilation and
extraction units

Art Equipment
3 Craven Street
Northampton
NN1 3EZ
tel. 01604 632447
Hotplates, etching presses, aquatint boxes, tailor
made acid trays, sinks and fume cupboards

Artichoke Print Workshop
Unit S1, Bizspace
245a Coldharbour Lane
London
SW9 8RR
tel. 020 7924 0600
www.printbin.co.uk
Suppliers of etching blankets and the KB
etching press

Mati Basis
13 Cranbourne Road
London
N10 2BT
tel. 020 8444 7833
Steel-facing copper plates

Senefelder
Andrew Purchess
6 Hyde Lane
Upper Beeding
West Sussex
BN4 3WJ
tel. 01903 814331

Regraining and supplier of zinc and aluminium
litho plates, intaglio suppliers and transport
of machinery

Smiths
London Metal Centre
42-56 Tottenham Road
London
N1 4BZ
tel. 020 7241 2430

Copper sheet

W&S Allely Ltd
PO Box 58
Alma Street
Smethwick
West Midlands
B66 2RP
tel. 0121 558 3301

Copper and aluminium sheet

John Pears
5 Witton Drive
Spennymoor
County Durham
DL16 6LU
tel. 01388 818004

Press engineer and manufacturer

Chris Holladay
Modbury Engineering
1, The Tanneries
East Street
Titchfield
Hampshire
PO14 4AR
tel. 01329 841000

Press engineer

Harry Rochat Ltd
15a Moxon Street
Barnet
Hertfordshire
EN5 5TS
tel. 020 8449 0023
www.harryrochat.com

Press engineer and manufacturer

Rollaco Engineering
72 Thornfield Road
Middlesborough
Cleveland
TS5 5BY
tel. 01642 813785
www.rollaco.co.uk

Press manufacturers. Suppliers of rollers, tools
and inks

Sericol
Westwood Road
Broadstairs
Kent
CT10 2PA
tel. 01843 866 668
www.sericol.co.uk

Suppliers of screenprinting equipment, inks and
sundries

Daler Rowney Ltd
Peacock Lane
Southern Industrial Estate
Bracknell
Berkshire
RG12 8ST

tel. 01344 424621
www.daler-rowney.com
Screenprinting system - base and acrylic colours

Hunter Penrose Ltd
32 Southwark Street
London
SE1 1TU
www.hunterpenrose.co.uk
Printmaking supplies and chemicals for etching
and lithography

Printmaking Today Magazine
Cello Press Ltd.
Office 18
Spinners Court
55 West End
Witney, Oxon
OX28 1NH
tel. 01993 701002
www.printmakingtoday.co.uk
Quarterly journal of contemporary international
printmaking; the authorized journal of the
Royal Society of Painter-Printmakers

Paintworks Ltd
99-101 Kingsland Road
Hackney
London E2 8AG
tel. 020 7729 7451
www.paintworks.biz
Conservation picture-framers; framing of artists'
prints a speciality

GLOSSARY OF TECHNICAL TERMS

The following list refers only to terms used in the body of this book. For more extensive technical information *see* Rosemary Simmons, *Dictionary of Printmaking Terms*, A&C Black, 2002. The present editors are indebted to the author for permission to quote in full or in part from her dictionary the starred items below.

Acetal resin: A synthetic resin material for engraving.

Aquatint: A method of achieving tone in an etching (*see below*) by sprinkling fine resin dust onto a metal plate, which is then heated so that the grains melt and adhere to it in tiny globules. On immersion in a mordant, the areas of exposed metal between the resin globules are etched by the mordant to create a texture which will hold ink.

Carborundum: Silicon carbide (SiC). An extremely hard abrasive widely used in print-making. Carborundum paper in various grit grades is used to smooth blocks and plates; as a grit, again in various sizes, it is used with a glue on a printing matrix to provide a tooth for printing ink. A carborundum print is an intaglio print characterised by the unusually heavy ink load which can be held by the grit on the plate.★

Chine collé: French term for thin paper, usually of Asian origin, which is glued, or collaged, on to a heavier backing sheet of paper. Often used only in the image area or in small separate areas on the print. Can be pre-printed and self-coloured on a colour different from the backing paper ...★

Collagraph: A print, intaglio and/or relief, taken from a collaged block or plate ... Materials such as cut-out shapes, found objects, organic substances can be glued to the base [plate] ...★

Computer print: See digital print.

Digital print: A print which has been created all or in part using digital equipment.★

Drypoint: A method of producing an intaglio print (*see below*) by simply scoring a plate (usually metal) with a steel point. The tool displaces (rather than removes) the metal, raising a ridge alongside each furrow. The ridge (called a burr) retains ink when the plate is wiped and prints a characteristically soft, velvety line.

Engraving: An engraving is a print made from a plate or a block engraved with a burin or graver rather than from a plate etched with a mordant.

Etching: Etchings are prints made from metal plates in which crevices are created through the action of a mordant (usually nitric acid, hydrochloric acid or ferric chloride). The image is drawn with a needle (or etching point) through a thin mordant-resistant coating of wax on the plate. The plate is then placed in a bath of the mordant so that the lines of the drawing are bitten into the metal. Both the extent and the depth of the biting can be controlled by successive immersions in the mordant, intermittently protecting with a mordant-resistant varnish those parts of the plate where biting is judged to have gone far enough. Etching plates are usually printed in intaglio (*see below*).

Flexograph: Lineblock or halftone made from rubber, plastic compounds or photopolymers, cheaper than metal. Widely used in commercial packaging printing but also used by artists …★

Inkjet printer: A computer printer which sprays tiny ink droplets onto the substrate. Bubble jet printers use heat expansion to create the pressure, piezo electrical charges drive others. They give near continuous tone quality. ★

Intaglio: Intaglio prints are made from (usually but not exclusively metal) plates engraved by hand or by the action of a mordant. The paper receives ink from the crevices of the plate rather than (as in relief prints) from the unengraved or unbitten surface. Ink is worked into the plate, which is subsequently wiped, leaving the crevices filled with the pigment. The plate is laid on the bed of a rolling press, face up; damp paper is placed on it, followed by several layers of woollen felt blanket, and the whole 'sandwich' is run through the press to produce the impression.

Linocut: Engraving on linoleum.

Lithography: A printing process depending on the incompatibility of grease and water. The method was perfected c.1798 in Bavaria by Alois Senefelder. The image is drawn with a greasy pigment onto a slab of absorbent stone (or onto a specially sensitized surface of lighter material). When the surface is sponged it all becomes wet except the drawn image, which repels the moisture. When greasy ink is rolled onto the damped surface it adheres only to the image. A press with a scraping action to apply pressure transfers the image to paper. The process has been developed to a sophisticated level for commercial use.

Mezzotint: An intaglio process in which the surface of a copper plate is evenly and systematically roughened all over with a rocker (a tool like a chisel with a short, broad blade whose cutting edge is curved and serrated). The dense texture raised by the rocker would, if inked and printed at that stage, produce a solid black. The full tonal range required in the final print is achieved by scraping and burnishing the textured surface to varying degrees.

Mixed media print: A print in which different printmaking techniques are used in combination.

Monoprint: An impression from a printing matrix which is uniquely different from other prints taken from the same matrix in terms of colour, paper or finish.★

Monotype: A single impression, not repeatable. A unique print where the image must be re-created each time, there is no re-usable printing matrix …★

Photo-etching: An etching which incorporates photographic images. Screenprinting, lithography and other methods are also capable of employing photography.

Pigmented inkjet: An inkjet print made with specially light-fast inks.

Relief print: A print made by applying ink and printing from the surface of a block (as distinct from pushing into its crevices, as in intaglio printing). The block's uncut surface is the area which is printed; the cut or indented areas show as the colour of the printing paper in the finished print.

Screenprint: A print made by supporting a stencil on a fine mesh stretched over a frame. The stencil may take the form of cut paper, adherent film or any ink-resistant pigment. It can also be created photographically.

Soft ground etching: Thin grease-proof paper is laid on a plate covered with a layer of soft mordant-resistant wax. Pencil lines or other marks that are then made on the paper will lift wax from the plate beneath when the paper is peeled off it. An image can in this way be produced in the wax and subsequently bitten into the plate by etching in the usual way.

Woodcut: A relief print made from a wood block engraved on the plank side; that is, with the grain.

Wood engraving: A relief print made from a wood block engraved on the end grain of a hard, close-grained wood such as box or lemon.

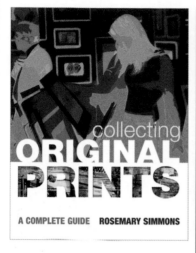
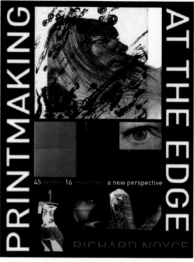

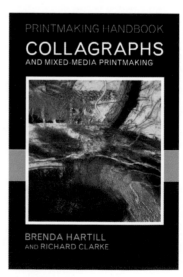

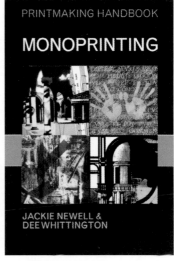
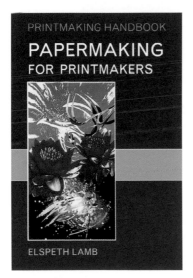